PROFESSIONAL DEVELOPMENT FOR TEACHERS OF MATHEMATICS

A HANDBOOK

Edited by
Ross Taylor

Author Team

Martin L. Johnson   Joan L. Akers
Jane D. Gawronski   Steven Leinwand
John A. Dossey   Ross Taylor
Shirley Frye   Leigh Childs

Advisory Committee

Ross Taylor   Peggy Neal
Carey Bolster   Thelma Sparks
Hilde Howden   Charles Hucka
Howard Johnson   Margaret J. Kenney
Steven Leinwand

National Council of Teachers of Mathematics
1906 Association Drive
Reston, Virginia 22091
National Council of Supervisors of Mathematics

Library of Congress Cataloging-in-Publication Data:

Professional development for teachers of mathematics.

    Bibliography: p.
    1. Mathematics--Study and teaching (Continuing
education) 2. Mathematics teachers--Training of.
I. Taylor, Ross.
QA11.P658   1986   510'.7'1   86-5332
ISBN 0-87353-231-7

Printed in the United States of America

CONTENTS

PREFACE

In the past decade, public concern about mathematics education in North America has moved from the issues of basic skills and minimum competency to concerns for excellence and higher-order thinking skills. As we have moved from being an industrial society to being an information society, there has been an increasing public awareness of the importance of communication skills and mathematical skills. Mathematics is also receiving greater support because of its importance in technology. The professional development of teachers is seen as a key to improving instruction in mathematics.

The interest we are seeing today in mathematics and science is comparable to the interest that surfaced after the launching of *Sputnik I* in 1957. At that time we felt threatened by international competition in space and military technology. Today we feel threatened by international economic and technological competition. Then and now, our response has included efforts to improve the teaching of science and mathematics. In the wake of *Sputnik I*, federal funds for the professional development of mathematics and science teachers flowed from the National Science Foundation to colleges and universities. These funds are beginning to be available again in a limited way. For example, under Public Law 98-377, the U.S. Department of Education provides funds that are administered through the states to the school districts, colleges, universities, and other agencies.

In light of the renewed interest in the professional development of mathematics teachers, the NCTM Professional Development and Status Advisory Committee prepared a position paper on *Professional Development Programs for Teachers of Mathematics*, endorsed by the NCTM Board of Directors in April 1985. (The position appears in Appendix 1 of this handbook.) The NCTM Board also approved and funded the development of this handbook to provide suggestions for implementing the recommendations of the position paper.

This handbook incorporates information on mathematics education and staff development that has become available since the publication of NCTM's *In-Service Handbook for Mathematics Education* in 1977. This handbook is designed to help educators address current issues in mathematics education such as the focus on problem solving, the use of manipulatives, the use of calculators and computers, achieving sex and race equity, the mathematics teacher shortage, and the teaching of new topics such as discrete mathematics, computer applications of mathematics, and probability and statistics.

The target audiences for this handbook are mathematics teachers, supervisors, administrators, college mathematics educators, and mathematicians. The objective of this handbook is to provide these persons practical ideas for designing and implementing professional development programs for teachers of mathematics.

ACKNOWLEDGMENTS

The Advisory Committee met in June 1985, identified the urgent need for this Handbook, and recommended that it be completed by 1 April 1986 in time for the NCTM Annual Meeting. The timely publication is a result of the splendid efforts of the Author Team, the Advisory Committee, and Charles Hucka and the NCTM staff. In addition, valuable suggestions and ideas were contributed by Henry Alder, Bob Baretta-Lorton, Carole Gupton, Marcia Horn, David R. Johnson, Harvey Keynes, Nancy Kreinberg, Charles Lamb, Calvin Long, John McBride, Bruce Meserve, Alan Osborne, and Linnea Weiland.

PART I
PROFESSIONAL DEVELOPMENT: AN OVERVIEW OF THE ISSUES

Professional development is the lifeblood of effective instruction in mathematics. What is professional development, why is it necessary, and what does research tell us?

In April 1985, the National Council of Teachers of Mathematics issued a position statement on *Professional Development Programs for Teachers of Mathematics* (see Appendix 1). That statement noted that

> because mathematics and education are disciplines that grow and change, teachers cannot depend on what they learned as undergraduates to carry them through their entire careers. Findings of research continually increase our understanding of teaching and learning. Further, social and technological changes increase the average citizen's need to understand and use mathematics. These forces demand reconsideration of the content and methods of mathematics instruction.

To address these realities and best serve the mathematical needs of students, teachers of mathematics require ongoing professional development to maintain and enhance their teaching skills and knowledge. Whether through workshops or courses, attendance at conferences, or collegial interaction, professional development is essential for effective mathematics instruction.

WHAT IS PROFESSIONAL DEVELOPMENT?

For this handbook, the terms *professional development, staff development,* and *in-service programs* are used synonymously to describe those *planned, ongoing, and systematic activities designed to maintain, enrich, or improve the skills, knowledge, and abilities needed by educational personnel to meet their professional responsibilities.* At all times, the ultimate goal of professional development is the improvement of student learning. Professional development encompasses a broad range of activities, from formal course work to the informal sharing of ideas with a colleague. It includes such activities as the following:

- Single workshops or a series of workshops led by fellow teachers, district-level mathematics specialists, or consultants

1

- Formal courses or seminars led by fellow teachers, district-level mathematics specialists, college or university personnel, or consultants
- Professional conferences, conventions, and seminars
- Residential courses or retreats
- Participation in formal or informal peer-group meetings to share ideas, problems, and information
- Visits to other schools
- Leaves of absence and sabbaticals
- Apprenticeships or internships in other organizations
- Evaluating textbooks
- Planning and developing curriculum
- Receiving informal on-the-job advice and assistance
- Being formally evaluated for the purpose of improving professional performance
- Observing demonstration classes
- Working in a teachers' center
- Individual research and planning focusing on the improvement of classroom instruction
- Self-teaching
- Self-evaluation for professional growth
- Teaching an experimental course

WHY IS PROFESSIONAL DEVELOPMENT NECESSARY?

The following goals for teachers, schools, and school districts were adapted from NCTM's *In-Service Handbook for Mathematics Education*, published in 1977. The imperative for high-quality professional development flows from these goals.

1. *To provide teachers the opportunity, the time, the means, and the materials for improving professional practice.* Curricular and instructional changes do not occur automatically. Rather, new ideas and techniques are integrated gradually into classroom practices, fostered by an individual and systemwide commitment to professional development. Improved professional practice, directed toward improved student learning, must be the focus of all professional development.

2. *To assist teachers in applying new understandings and insights in the learning process.* The art and science of teaching derive from a blend of content and method. Results of research in different aspects of the learning process hold significant implications for teaching practice that teachers

should know. Teachers need assistance in learning to apply these new insights to their own teaching methods, especially when changes in customary procedures may be required. It is generally recognized that a gap of decades exists between research findings and classroom practice. An important purpose of professional development is to reduce this gap. Concurrently, changing societal needs and demands require the periodic review and adjustment of mathematical content. New content, such as computer applications or discrete mathematics topics, can be effectively implemented only after teachers have learned the content and have suitable teaching materials.

3. *To help teachers expand their perceptions of mathematics.* Teachers of mathematics often look on their subject as merely a necessary "tool." They may justifiably lack an appreciation of the intellectual richness, cultural significance, and innate beauty of mathematics, its logical structure, and its varied processes. Because teachers may be the product of uninspired mathematics teaching in their own education, they may convey to their students a similar lack of enthusiasm for, and satisfaction in, learning and using mathematics. A range of professional development opportunities can help greatly in broadening teachers' perceptions of mathematics and its importance.

4. *To assist teachers in their personal development as professionals.* Teachers should see themselves as professionals rather than just employees. Caring about teaching goes hand in hand with sharing with colleagues. Teachers should experience the professional stimulation derived from such activities as reading and contributing to professional journals, attending and making presentations at professional meetings, and serving as officers or committee members of professional organizations. Teachers should be aware that NCTM and its affiliated groups offer them a wealth of professional opportunities. They should feel the true satisfaction of translating stimulating professional experiences into motivating classroom activities.

5. *To assist teachers in developing creative instructional approaches that are meaningful and mathematically correct and that instill in students enthusiasm and satisfaction in learning and using mathematics.* Goals 1, 2, 3, and 4 flow naturally into goal 5—improved classroom practice that maximizes student interest and learning.

In addition to furthering goals that address individual teachers, professional development should meet broader systemwide goals:

6. *To maintain quality within the existing mathematics program.* For an educational program of high quality to continue to function well, there should exist a means for the frequent review of curricular guidelines and the updating of both content and pedagogy. Seminars for members of a mathematics department or discussion groups for teachers of specific courses or grade levels are excellent professional development opportunities.

7. *To provide a mechanism for responding to problems of a curricular or instructional nature*. Test results, parental concerns, new developments, and a host of other sources of information often reveal problems or needs within a mathematics program. Professional development programs such as demonstrations, workshops on new materials, and teacher visitations are ideal vehicles for effectively addressing problems and implementing solutions.

8. *To implement significant and innovative curricular and instructional practices*. There is room for improvement in all mathematics programs. Courses can be updated, curriculum expanded and sequenced, and methods adopted that reflect recent research findings. Regardless of the type or degree of change, professional development is a necessary component of effectively implementing any change. The greater the changes expected in teaching behavior, the greater the need for professional development. Significant changes should be supported with extended professional development programs that allow the teachers time to assimilate the new ideas and put them into practice. Professional development should be accomplished through a comprehensive, ongoing plan.

WHAT DOES RESEARCH TELL US ABOUT STAFF DEVELOPMENT?

The following summary of highlights from research on staff development for effective teaching appeared in the November 1983 issue of *Educational Leadership*:

> Studies comparing various models for processes of staff development are rare. While it is not possible to state conclusively that one in-service design is superior to another, we can put together . . . many pieces of development programs for more effective teaching.
>
> 1. Select content that has been verified by research to improve student achievement.
>
> 2. Create a context of acceptance by involving teachers in decision making and providing both logistical and psychological administrative support.
>
> 3. Conduct training sessions (more than one) two or three weeks apart.
>
> 4. Include presentation, demonstration, practice, and feedback as workshop activities.
>
> 5. During training sessions, provide opportunities for small-group discussions of the application of new practices and sharing of ideas and concerns about effective instruction.
>
> 6. Between workshops, encourage teachers to visit each others' classrooms, preferably with a simple, objective, student-centered observation instrument. Provide opportunities for discussions of the observation.
>
> 7. Develop in teachers a philosophical acceptance of the new practices by presenting research and a rationale for the effectiveness of the techniques.

Allow teachers to express doubts about or objections to the recommended methods in the small group. Let the other teachers convince the resisting teacher of the usefulness of the practices through "testimonies" of their use and effectiveness.

8. Lower teachers' perception of the cost of adopting a new practice through detailed discussions of the "nuts and bolts" of using the technique and teacher sharing of experiences with the technique.

9. Help teachers grow in their self-confidence and competence through encouraging them to try only one or two new practices after each workshop. Diagnosis of teacher strengths and weaknesses can help the trainer suggest changes that are likely to be successful—and, thus, reinforce future efforts to change.

10. For teaching practices that require very complex thinking skills, plan to take more time, provide more practice, and consider activities that develop conceptual flexibility.

A comprehensive research-based approach to designing and implementing effective professional development incorporates the following five stages:

- Readiness
- Planning
- Training
- Implementation
- Maintenance

Appendix 2 contains a list of the ten basic assumptions on which these Readiness, Planning, Training, Implementation, Maintenance (RPTIM) Model Practices are based and a listing of thirty-eight specific practices.

CHARACTERISTICS OF EFFECTIVE PROFESSIONAL DEVELOPMENT PROGRAMS

The characteristics listed below are adapted from the National Staff Development Council's twenty "Characteristics of Effective Staff Development Activities":[1]

1. *Involvement in Planning Objectives*
Staff development activities tend to be more effective when participants have taken part in planning the objectives and the activities. Objectives planned by the participants are perceived as clearer and more meaningful and thus have a higher degree of acceptance.

---

1. The 20 Characteristics of Effective Staff Development Activitiesappeared in a National Staff Development Council Manual forConducting Effective Staff Development Programs.

2. *Active Involvement of the Building Principal*

Staff development activities in which the building principal is an active participant have proved over a period of time to be more effective.

3. *Time for Planning*

Whether the staff development activities are mandated or participation is voluntary, participants need time away from their regular teaching or administrative responsibilities to plan objectives and subsequent activities.

4. *District Administrative Support*

For activities to be effective, district-level support needs to be visible.

5. *Expectations*

Participants should know what they will be able to do when the experience is over and how the experience will be evaluated. They should also know what will be expected of them during the activities.

6. *Opportunity for Sharing*

Activities in which participants share and assist one another are more apt to attain their objectives than activities in which participants work alone.

7. *Continuity*

Activities that are thematic and linked to a staff development plan or a general effort of a school are usually more effective than a series of one-shot approaches on a variety of topics.

8. *Expressed Needs*

Effective activities are based on a continuous assessment of participants' needs; as needs change, the activities should be adjusted accordingly.

9. *Opportunity for Follow-up*

Activities are more successful if participants know there is an opportunity to become involved in follow-up sessions.

10. *Opportunity for Practice*

Activities that include demonstrations, supervised tasks, and feedback are more likely to accomplish their objectives than those that expect participants to store up ideas and skills for use at a future time.

11. *Active Involvement*

Successful staff development activities give the participant a chance to be actively involved. When hands-on experiences with materials, active participation in exercises that will later be used with students, and involvement in small-group discussions are used, participants are more likely to apply what they have learned when they return to their regular assignments.

12. *Opportunity for Choice*

If a participant has chosen to become involved in an activity, there is a far greater likelihood that the experience will be meaningful. A meaningful series of alternative activities should be offered within a staff development program that is planned over a period of time.

13. *Building on Strengths*

People like to be recognized as valued, competent, liked, and needed. Staff development activities that view each participant as a resource are often more responsive to participants' needs.

14. *Content*

Successful staff development activities appear to be those that are geared toward a relatively narrow grade-level range, a specific topic, a specific set of skills, a plan that is ready for immediate use when the participant leaves the activity, or a set of instructional materials that translate the ideas into practice.

15. *The Presenter*

Successful staff development activities are those in which the presenter has been able to approach the subject from the participant's point of view. The presenter's expertise also plays a role, as does his or her ability to convey genuine enthusiasm for the subject.

16. *Individualization*

Staff development activities that have different educational experiences for participants at different stages of their development are more apt to attain their objectives than programs in which all participants engage in common activities.

17. *Number of Participants*

Some presentations are as effective with 100 participants as they are with ten. However, for staff development activities requiring personal contact, informality, and an interchange of ideas, seven to ten participants appear to be optimal. There are exceptions and variations based on the skill of the presenter, the organization of the activity, and the nature of the topic.

18. *The Learning Environment*

As a general rule, more successful staff development activities take place within a low-threat, comfortable setting in which there is a degree of "psychological safety." Openness to learning appears to be enhanced when peers can share similar concerns, highs and lows, problems and solutions.

19. *The Physical Facility*

Accessibility of supporting materials, appearance of the facility, room temperature, lighting, auditory and visual distractions, and many other physical factors have subtle, but sometimes profound, effects on the success of the staff development activity.

20. *Time of Day and Season*

Staff development activities that take place at the end of a school day have less chance of being successful than those offered when participants are fresh. Further, staff development activities are less likely to be successful when they are scheduled at times of the year when seasonal activities, parent conferences, holiday celebrations, and so forth, occur.

WHAT ARE THE KEY ISSUES FACING TEACHERS OF MATHEMATICS?

Many issues face mathematics educators today. These issues should be considered, and their relevance to the local curriculum examined, before planning a professional development program. These are among the important questions mathematics educators face:

1. How can the mathematics curriculum be truly organized around problem solving?

2. What experiences, such as the use of concrete materials, promote the understanding of mathematical concepts?

3. How can students best develop the "number sense" essential for algorithmic understanding, estimation, mental arithmetic, and determining the reasonableness of results?

4. How does having students verbalize, through writing and sharing ideas in cooperative learning groups, help clarify their mathematical thinking?

5. How can calculators be effectively incorporated into the mathematics curriculum?

6. How can computers be used to enhance the mathematics curriculum?

7. How can we ensure equal access to calculators and computers for all students?

8. How will the impact of computing technology affect the amount of time devoted to developing proficiency with multidigit algorithms?

9. What effect should the pervasive use of statistics in our society have on expanding the amount of time spent on data collection and analysis in the mathematics curriculum?

10. How can teachers help students see the relevance of mathematics to their world?

11. How can teachers help underrepresented groups (females and minorities) realize their full potential in mathematics?

12. How can teachers establish a learning environment where mathematics anxiety is unknown?

13. How can teachers best prepare college-bound students for the college programs they will enter?

14. How can the shortage of fully qualified teachers of mathematics be addressed?

In addition, new information about general teaching techniques has come to light in recent years. Some examples follow:

- Cooperative learning
- Higher student expectations
- Teaching to objectives
- Use of effective teaching models
- Increasing time on task
- Effective questioning techniques

Professional development is the key to implementing the best practices currently known in the classroom. Professional development is a force that drives other forces. (For example, textbooks have a strong influence on instruction. Textbooks are selected by teachers. Well-informed teachers will select better textbooks and will influence publishers to continue to improve their products.) Professional development programs need to inform teachers about promising practices and provide experiences that will enable the teachers to implement these practices in their classrooms.

THREE IMPORTANT FEATURES OF PROFESSIONAL DEVELOPMENT PROGRAMS

Teachers of mathematics, like other professionals, require ongoing cumulative professional development. Research findings have increased our understanding of the teaching/learning process. Societal changes have increased the need for mathematical knowledge. As a result, students need instruction that combines content and pedagogy, using effective learning materials. To accomplish this, teachers of mathematics need professional development experiences that artfully blend content knowledge, instructional pedagogy, and resource materials.

KNOWLEDGE OF CONTENT

The old adage "A good teacher can teach anything" simply does not hold up when it comes to teaching mathematics. To be a good teacher of mathematics, one must know some mathematics. Although many teachers learned a great deal of mathematics in college, there is always a need to learn more, or at least to learn new things about previously learned topics. If one believes that education is a lifelong enterprise, then it is crucial that all teachers of mathematics continue to be students of mathematics. To this end, programs of professional development for mathematics teachers should include a reasonable amount of time devoted to the acquisition of content knowledge that is new to them.

PEDAGOGY

Besides knowing what to teach, mathematics teachers must also have the skills of knowing how to teach. Professional development programs should include a reasonable amount of time for methodology. Whether teaching concepts, skills, or applications, teachers should have the opportunity to become acquainted with effective teaching practices that have been identified through research and experience. Then, teachers will have the potential to deal with such pragmatic concerns as using technology in teaching mathematics, addressing individual differences, and developing understanding of mathematical concepts.

RESOURCE MATERIALS

In our fast-paced society, new technology and new materials are continuously being developed. With the day-to-day demands placed on classroom teachers, it is virtually impossible for them to keep abreast of all the new developments in mathematics education. Therefore, it is extremely important to allow time for teachers to make, develop, take, or at least become aware of sources of resource materials for the mathematics classroom. In short, "give them something to take home," and include "something they can use tomorrow." In addition, give them something—perhaps a provocative article—to make them think critically about their teaching and make them want to get back together for more sharing and more professional growth.

PART II
GUIDELINES FOR EFFECTIVE PROFESSIONAL DEVELOPMENT ACTIVITIES

Part II is divided into five sections:

- Building Support and Commitment
- Planning the Program
- Choosing a Format
- Implementing the Program
- Evaluating the Program

Each of these sections corresponds to one of the five guidelines set forth in the NCTM position statement on *Professional Development Programs for Teachers of Mathematics*. The statement of the corresponding guideline appears in a box at the beginning of each section.

BUILDING SUPPORT AND COMMITMENT

> Professional development programs for teachers of mathematics should be based on a strong commitment to professional growth.
>
> *a*. An appropriate person should be responsible and accountable for the professional development of the teachers.
> *b*. Sufficient time should be allocated for individuals to assess needs, plan activities, lead or participate in programs, and evaluate outcomes.
> *c*. Sufficient funds should be available to support professional development programs and ensure teachers' participation in them.

THE ROLES OF THOSE WHO ARE RESPONSIBLE

As noted in the research findings discussed in Part I, successful professional development is dependent on the active support and participation of many individuals. In particular, professional development provides the greatest impact on mathematics classroom instruction when it is a cooperative venture that involves teachers, administrators, supervisors of mathe-

11

matics, college mathematics educators, and mathematicians. Each contributes a vital component and fulfills a specific role of responsibility.

Teachers should be aware of the benefits of professional development and should seek out opportunities that will help them to grow professionally. Even when opportunities are limited, teachers can promote professional growth for themselves and their colleagues by availing themselves of the resources of NCTM and its affiliated groups, by demonstrating leadership in their school districts, and by working through their professional organizations to obtain school policies that are favorable to professional development.

Administrators should be aware of current research in effective staff development. They should assess professional growth needs and identify leaders who can provide staff development in mathematics. They should develop a climate in which professional development is encouraged and rewarded by arranging for substitute service, providing funding for professional activities, and giving recognition to teachers for their professional contributions.

Supervisors of mathematics should be ultimately responsible and accountable for professional development opportunities for mathematics teachers. In order to fulfill these obligations, they should be familiar with strategies of effective mathematics teaching and staff development procedures, know the implications on curriculum of current local and national educational reform movements, and be provided with sufficient time and resources. In districts that do not have a supervisor of mathematics, a qualified staff member, such as a secondary school mathematics department head, should be designated to fulfill the role and be provided with training and sufficient released time and resources to assume the necessary responsibilities.

College mathematics educators are an integral part of their mathematics community. They should participate in the activities of NCTM and its local affiliated group, encourage such participation of their students, and provide leadership in coordinating professional development opportunities for teachers. They can work cooperatively with school districts and professional organizations to develop and provide in-service courses. They can seek government and private funding for staff development programs.

Mathematicians should be the most outspoken supporters of professional growth opportunities for teachers of mathematics. It is they who should be most aware of the changing role of mathematics in society and the rapidly growing need for mathematical skills in business, industry, and government, and it is they who can provide leadership in the curriculum changes necessary to meet this growing need. Their renewed interest in precollege mathematics eduction can have a significant impact on upgrading the quality of mathematics instruction in the elementary and secondary schools. Mathematicians who become involved in the professional development of teachers should be familiar with the classroom environment in which teachers function.

INCENTIVES FOR PROFESSIONAL DEVELOPMENT

One clear form of district- or school-level support for professional development is the provision of incentives. Persons involved in the professional development of teachers need to be fully aware of the incentives that motivate teachers to participate. Incentives should be part of the design of any professional development program, and publicity to encourage participation should feature the incentives.

Released time is an incentive. Teachers are often involved in many activities such as supplementary jobs, coaching, supervising extracurricular activities, church and community activities, recreational activities, and family responsibilities. Time for professional development competes with time for these activities. Many school districts address this problem by providing released time for professional development activities. Either certain days or parts of days are scheduled for staff development activities and the students are released from school or substitute service is provided so that teachers can participate in professional activities.

Disadvantages of released time are the decrease in time the teacher spends with students and the expense of hiring substitutes. Advantages include the affirmation that professional development is important, the lack of conflict with outside activities, and the opportunity for the teachers to concentrate on professional development in a relaxed environment.

Financial incentives generally fall into two categories. Teachers may be reimbursed for the cost incurred in participating in workshops, courses, or conferences, or they may be paid a stipend to cover time spent in approved staff development activities outside the school day.

Credit incentives include college credit, credit within the district toward promotion or salary increases, and credit required for recertification. Such credit may be awarded in lieu of financial incentives or be combined with financial incentives when a local system or an external funding source defrays part or all of the costs of earning credit.

Additional incentives include social functions, such as dinners or retreats, free materials for use in the classroom, public recognition, and letters of commendation placed in the teacher's personnel file.

Intrinsic incentives appeal to the desire of teachers to do a better job of teaching. These incentives include—

- the opportunity to learn new ideas;
- the opportunity to share ideas and experiences;
- intellectual stimulation;

- the opportunity to network with colleagues;
- professional recognition by peers.

These intrinsic incentives can counteract the forces in education that tend to isolate teachers from their colleagues. Teachers can participate collaboratively in a rich and rewarding profession instead of viewing teaching as a lonely occupation.

FORCES THAT IMPACT ON PROFESSIONAL DEVELOPMENT

Anyone who has ever used a list of pros and another list of cons to help in making a decision has used a form of Kurt Lewin's force field analysis technique, which examines the driving forces and the restraining forces that are present in any situation that involves change. For the change to occur, the sum of the driving forces must be greater than the sum of the restraining forces. Since support for professional development results from a recognized need for some form of change, Lewin's technique aids in analyzing the forces at work.

Every professional development activity is impelled by a variety of forces seeking to enact change. Unfortunately, these driving forces are usually accompanied by a variety of restraining forces, as illustrated in figure 1.

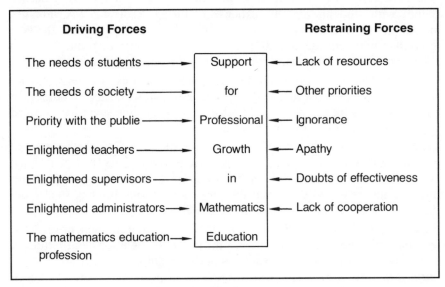

Fig. 1

Persons who wish to obtain support for professional growth in mathematics education should be aware of these forces and act to maximize the driving forces and minimize the restraining forces.

PLANNING THE PROGRAM

Professional development programs for teachers of mathematics should be carefully planned.

a. Clear objectives should be established.

b. The programs should improve students' learning experiences by improving the skills and knowledge of their teachers.

c. Those whom the programs are designed to assist should contribute significantly in planning the programs.

d. Extensive assessments of individual and collective needs should serve as bases of the programs.

e. Current concerns and issues in mathematics education should be reflected in the content of the programs.

f. The programs should be ongoing and cumulative.

High-quality staff development programs are essential for high-quality mathematics instruction. To be effective, such programs need careful planning and preparation. Since implementing change requires a long-term commitment and a realization that the ideal is never attained but continually pursued, professional development for mathematics teachers should be ongoing and cumulative.

ASSESSING NEEDS

A needs assessment helps determine the status quo and provide direction. Information from needs assessments can be used to identify perceived strengths and weaknesses, evaluate the impact of a specific program, set priorities, and determine the next steps. Three areas of concern could be addressed in the needs assessment:

1. How appropriate is the mathematics *curriculum* for the students, and how well does it reflect current thinking and research in mathematics education?

2. Are the *teachers* sufficiently knowledgeable of the content and are they using appropriate teaching strategies to deliver the curriculum?

3. What *support* do teachers have in implementing the curriculum, both personally and with appropriate materials?

In reality, much of a needs assessment can be accomplished informally. Teachers want to keep current on issues in mathematics education; administrators feel that test scores in mathematics are too low; a new textbook or program has been purchased and teachers need training in its use; institutions of higher education receive a grant and seek districts, schools, and teachers to participate. Although informal procedures are often used, more formal methods of assessing needs may help ensure the success of a professional development program.

In general, the methods used should correspond to the extensiveness of the staff development program to be planned. Needs assessment procedures that require a significant amount of the teachers' time will be ineffective unless the teachers see a real payoff in a well-planned program that meets their expressed needs. Needs assessments that incorporate a variety of measures can present a clear picture of what currently exists and what could exist. The following are possible ways to collect information for assessing needs.

1. *Analyze existing documents.* The existing local curriculum or textbooks can be compared to district, state or provincial, or national standards or recommendations. Examining such information as test results, prior needs assessments, and goal statements can also be a sensible starting point. This information should be readily available and may highlight areas that need further investigation.

2. *Interview staff.* Much useful information can be gathered through informal conversations. Structured interviews, however, provide the opportunity to collect comparable data from a variety of persons. A representative sampling should include all groups to be affected by the staff development program, including principals and other administrators, as well as teachers.

3. *Observe classroom practices.* Classroom observation provides information on what is actually occurring in the classrooms. If the observations are conducted by those responsible for the planning or presenting of the inservice program, the program can be tailored to the needs of the participating teachers. Teachers should be assured that information gathered through observation as part of a needs assessment will not be used in teacher evaluations.

4. *Conduct surveys.* Surveys and questionnaires are often viewed as the primary source of needs assessment data. Since it takes a great deal of time to design and analyze surveys that will provide appropriate information, this option should be considered when other methods do not provide adequate data. Surveys and questionnaires provide baseline data that can be used to show growth that occurs as a result of the staff development program. (See page 41 in the Evaluation section.)

5. *Facilitate group discussions.* Group discussions allow participants to pool their ideas and collaboratively identify staff development needs. This approach can stimulate active involvement in the proposed in-service programs.

6. *Employ outside consultants*. Outside consultants can frequently bring new insights and perspectives to help schools and districts conduct needs assessments and plan innovative, appropriate professional development programs.

DEVELOPING A PLAN

Effective professional development requires a strong and visible long-term commitment by the school or district. A one-shot in-service program held in isolation from other professional development activities will have little lasting effect on teaching practices. Improvement in mathematics instruction will happen only when teachers and administrators—

- become aware of a need for change;
- become knowledgeable of alternative methods of effectively teaching mathematics;
- have opportunities to practice and incorporate identified methods in their classrooms through a support network or coaching.

The development of a long-term school or district professional development plan will help assure that suggested improvements in the mathematics curriculum carry over to classroom implementation.

Responsibility

A key element in planning a successful staff development program is identifying one person to be responsible and accountable for the planning and implementation of the program. This person should be knowledgeable about mathematics education and have the time and authority to coordinate the planning and implementation of the professional development plan.

Planning Committee

Although one person should assume major responsibility for coordinating the professional development program, the planning committee should represent program participants to ensure a successful program. This group will be able to contribute many ideas to help ensure that the needs of the mathematics teachers are being met. It may also be beneficial to include as a committee member an outside expert in mathematics education who can bring a broader perspective on issues to be addressed.

Goals and Objectives

The purpose of the planning committee is to establish the general direction and specific focus for the professional development program. The program is more likely to be effective if it concentrates on a particular theme. An early task is to set overall goals and objectives. These goals and objectives should be based on the results of the needs assessment and related to the critical issues that face mathematics educators. The content, scope, and format of the program should reflect these goals and objectives. The objectives should be stated so that all those involved (participants as well as planners) have a clear picture of the intended outcomes. The intended outcomes should be viewed in terms of expected changes in the behaviors of administrators, teachers, and students.

Action Plan

The planning committee needs to develop a long-range action plan and determine the scope of the professional development program. The detail and complexity of the plan will depend on the size and resources of the school or district. Questions that need to be addressed in the development of the long-range plan include the following:

1. *Audience.* Is the professional development program mandated for all teachers of mathematics classes in a school or district, or is the program intended for a select group? Should two or three persons from each school participate and then be expected to share what they learn with their school staffs? Should participants include administrators, department chairs, or aides whose support will be necessary for successful implementation at a school site?

2. *Format.* What type(s) of programs will be the most effective? Hands-on workshops? Ongoing school study and discussion groups? Enrollment in college courses or institutes? Conference attendance? Or a combination of activities? (See the section on choosing a format beginning on page 22.)

3. *Scheduling.* How often should in-service activities be scheduled? Will there be time for teachers to try out ideas with their classes and report their successes and failures to other participants? What in-classroom support—for example, peer coaching—will teachers receive? Is the program spread out over a time period long enough to allow teachers to become comfortable with new practices?

4. *Budget.* Is allocated funding sufficient, or will additional support be needed? (Suggestions for procuring outside funding appear in Part III: Finding Resources.) Some budgetary questions to consider are listed below.

- *Teacher release.* Will the program be scheduled during the school day and require funding for substitute teachers, on nonschool days requiring funding for teacher stipends, or before or after school?

- *Staff developers.* Will outside consultants, who may require fees and travel expenses, conduct the activity, or will district staff be enlisted?

- *Materials.* Will materials be necessary for the program or for successful implementation in the classroom?

- *Refreshments*. What refreshments will be served and how will they be financed?

When the committee is ready to develop short-term plans, they will need to determine specifics, such as the number and frequency of sessions, when and where they will be held, who should attend, and how the program will be evaluated.

PLANNING THE SPECIFICS

Adequate preparation time is crucial. Time is needed to select the leader(s), determine his or her availability, arrange a location, and give participating teachers adequate notice. If time allows, it is advantageous for the coordinator or leader to visit each participant ahead of time to get to know them and their specific needs.

In-Service Leaders

Great care should be taken to select the best possible session leader(s). This is probably the most important decision to be made. If the committee does not know a presenter, it should try to verify recommendations by checking with several people who do. The objectives of the in-service program and the expectations for the individual sessions must be clearly communicated to the session leaders.

Using two leaders has several advantages. Each can assist the other with materials, audiovisual equipment, and any problems that arise. Two leaders can assess better the group's response and adjust the focus to meet the participants' needs. Participants will benefit from the expertise of two individuals.

On-Site Coordinator

A person other than the session leader(s) should be assigned the task of coordinating the program. Responsibilities include notifying the participants, making certain the room is set up, refreshments are ready and replenished, handouts and materials are prepared, and name tags are made. The coordinator is available to help with handouts and audiovisual aids. It is understood that the coordinator will handle any emergencies that might arise, such as a cold room, malfunctioning equipment, a shortage of coffee, or the cancellation of a speaker.

Location

Many in-service programs for educators take place at school sites. If circumstances allow meeting at a district office, hotel, business, college, or university, try to select a central site with adequate parking. When selecting the room(s), consider physical comforts; these can contribute significantly to the success of the event. Make certain the furniture is comfortable and the temperature-control capabilities are adequate. Space should be available for breaking into groups. A sink or a place to brew coffee is helpful.

Setup

A table arrangement is preferable to rows of chairs for most professional development sessions. Tables encourage interaction and the use of hands-on materials to involve the participants.

The coordinator should inquire in advance about the audiovisual needs of the presenters. An overhead projector and flip charts will keep the audience's attention and help illustrate points. Many quality videotapes are available to enhance presentations. Some presenters may need computers, large-screen monitors, or other equipment. Thus the placement of outlets may be a factor in the setup.

Publicity

Effectively publicizing professional development events will ensure a successful beginning. With adequate advance notice, schools and participants will have time to make the necessary arrangements to attend. The initial announcement should include the following:

- Type of training and its purpose
- Who, when, where
- Financial obligation (if any)
- Notice of subsequent sessions
- Deadline for responding
- Address for responding
- Phone number for questions
- Response form

A reminder or confirming notice to participants should include this information:

- Beginning and ending times
- Arrangements for meals (if applicable)
- Directions to in-service site (if appropriate)
- Schedule of events
- Expected outcomes

Agenda and Evaluation

An agenda with specified times should be developed with the in-service leader(s). This should be duplicated for the participants or posted where it can easily be seen. An evaluation form should be prepared to assess how well the goals and objectives are met. A sample evaluation form appears on page 65.

Refreshments

Provide plenty of coffee, including decaffeinated, as well as hot water and tea bags. Doughnut holes, fruit, cookies, or other refreshments can provide needed energy in the morning or a pick-me-up after a day of teaching. On occasion punch or other cold drinks may be appropriate.

If the session is a full day's event, teachers respond positively to going out for lunch. Getting away for lunch encourages discussions of the morning's ideas and the formation of informal teacher networks. It is helpful to provide a list of nearby restaurants, including directions and sample menus. Make arrangements so that no one is left to lunch alone. The restaurant crowds can be avoided by dismissing the group before noon. Another possible arrangement is to have an in-house lunch. Be sure to inform participants in advance of the arrangements for lunch.

Donations for refreshments can often be obtained from community groups, such as the PTA, local banks, or other businesses. For example, a local parent group could be asked to donate a salad bar luncheon. Staff developers can provide food and use the "kitty" method for reimbursement.

Handouts and Materials

Handouts enhance participation and facilitate the implementation of ideas. A file folder can be presented to each participant for keeping handouts; such a folder is ready to file for easy access. Label each folder with the event's name. Include in it such items as an agenda, goals and objectives, an evaluation form, a list of participants, and blank paper for taking notes.

Display recommended resources for participants to examine, and if possibleallow them to borrow the materials. If this opportunity is provided, have a checkout system to monitor the return of all borrowed items. When many resources are mentioned, it is helpful to distribute bibliographies and ordering information. With advance notice, publishers will gladly provide catalogs.

Attractive name tags, prepared in advance, also make the participants feel special. Networking can be enhanced by including such information as school and grade level.

A sample checklist of supplies that can help a coordinator prepare for a professional development event appears in Appendix 3.

CHOOSING A FORMAT

Professional development programs for teachers of mathematics should recognize individual differences.

  a. Varied formats, including workshops, conferences, institutes, courses, and in-school discussion sessions, should be used.

  b. Programs should be tailored to meet the needs of teachers whose knowledge, skills, and experiences are diverse.

The development of opportunities for professional growth that meet the identified needs of teachers and school programs is a challenging task that must reflect a multitude of factors, such as differences in participants' experience, knowledge, and pedagogical skills. Too frequently, professional development programs are limited to the college-lecture style of courses, extended workshops, or one-day "hypes." However, there are many alternative formats for in-service activities that have proved successful. These alternative formats for structuring professional development activities include possibilities for short- and long-term programs for large numbers of teachers, activities for small groups, and choices for individual teachers.

Short-Term Format

- Single-day conferences
- Dinner meetings
- Miniconferences on a single topic
- Professional organization conferences
- Workshops (from make-it-take-it to curriculum)
- Staff retreats
- Professional study days

Longer-Term Format

- Specific institutes (coordinated worksessions)
- Formal course work
- Course work combined with follow-up implementation

Small-Group Format

- Locally funded projects
- Reading groups
- Local school departmental activities

Individual Format

- Visiting other classes; being visited
- Sabbatical programs
- Individual reading/study programs
- Internships
- Conference participation

There are other alternatives that could be added to this list, but let us consider some of these options and the possibilities they hold.

SHORT-TERM FORMAT

Short-term programs can consist of a one-day or part of a day session, a two- or three-day conference, or a series of several short weekly or monthly sessions. There are limitations on what can be accomplished in a short meeting, but the format is well suited to achieve certain objectives:

- Alerting teachers to a new topic, a new approach, or a new set of materials
- Helping teachers reach a minimal level of performance in a new area
- Creating interest in a topic that will serve as the basis for a more thorough study later
- Introducing a new curriculum or branch of the curriculum

Well-planned professional development programs that require only a short commitment of teacher time tend to have a good probability of success. They not only create a feeling of ownership of the ideas considered in the program but also encourage their immediate use in the classroom. This continuity may be missing in long-term programs such as courses.

The following are examples of potential topics that could be used in one-shot ventures or minicourses with a specific purpose:

- Sequencing numeration materials for introducing place value
- Developing instructional strategies for learning basic facts
- Evaluating problem-solving activities in middle grade mathematics
- Analyzing time usage in mathematics instruction in the middle grades
- Developing models for rational number concepts
- Reviewing software for developing estimation skills
- Developing students' conceptions of equations
- Troubleshooting common errors in first-year algebra
- Introducing Gaussian elimination and systems of equations
- Applying combinatorial problem solving in geometry

- Introducing box-and-whisker plots in elementary statistics
- Preparing students for mathematics contests

Although some of these topics can easily be covered in a short series or in the two- to four-hour time period of a half-day meeting, others clearly require more time. These topics can be brought to an elementary level of understanding and serve as a springboard to longer, more involved professional development sessions. For example, the Gaussian elimination and combinatorial problem-solving sessions might each serve as attractors for an introductory course in discrete mathematics for secondary school teachers.

Such topics might also be presented to teachers through a series of dinner meetings or through individual reading or study materials developed in a central district office. In the former, teachers are served a meal in a professional and social atmosphere. In the latter, they are given more flexible time constraints for study.

An effective format is a miniconference constructed around a specific topic or theme and designed to involve teachers in the ties between learning, curriculum, and teaching. These meetings might be organized around the following framework:

1. Definition of the problem/topic
   a) Establish a specific need.
   b) Give a specific statement of the topic and its location in the curriculum.
   c) State specific goals for the session or series.
2. Knowledge of the topic from a research/theoretical standpoint
   a) Develop known results concerning a topic from mathematics education research through models, sequences, or developmental patterns.
   b) Develop a mathematical topic from known results tied to current topics in the curriculum, specific definitions, and theorems needed.
   c) Develop ways in which the topic can be integrated into the classroom. Specific attention must be given to timing, sequencing, pacing, problem areas, levels of understanding, applications, and assessment.
   d) Develop specific activities for classroom use in ready-to-use condition. These should be classroom tested and contain an instructor's key and a list of ready references for further study.

This format allows teachers to build a foundation, to see the relationship between the foundation and practice, and to be ready to apply the activities in their own classrooms.

An example of such a miniconference might be holding a four-hour program on classification for early childhood teachers.

> The session might begin with a consideration of classification and its roles in both mathematics learning and language development. A brief survey of developmental sequences related to classification could follow. The focus could then shift to specific activities related to the various developmental stages. Two activities could then be provided for each of the stages and time allowed for participants to work through these activities. Finally, the teachers might be given a list of extra activities related to those practiced in the session as well as a list of alternative readings and sources of ideas.

Another short-term form of professional development is attendance at professional meetings, such as the regional and annual meetings of the National Council of Teachers of Mathematics. Through these meetings, a school district can make a wide variety of opportunities available for their staff members. Teachers can be divided into different task forces to attend sessions on different topics and then work up a minisession for other teachers when they return. Timely topics can be pursued further through readings from the *Mathematics Teacher*, the *Arithmetic Teacher*, other NCTM publications, and additional sources.

Attendance at professional meetings, such as those of the NCTM and other mathematics organizations, does more than allow the teachers to passively hear ideas. They can participate in workshops, examine the latest in curricular materials, and exchange ideas with teachers from other schools who are having the same problems and experiencing the same changes in students and conditions. These informal conversations are an important part of professional development. Too often, professional development programs are viewed as a downward transmission of knowledge rather than a sharing among practitioners. The meetings of the NCTM and other professional groups are an excellent chance for teachers to share their successes and concerns in a professional atmosphere.

Districts can also offer a series of short workshops focused on specific materials or points in the curriculum at a given grade level. These workshops should range from "make it, take it" in format to learning the uses of a specific teaching aid. When teachers have an opportunity to see and work with materials, they are more likely to use those materials in their own classrooms. The opportunity to practice with the materials, get feedback on that practice, and then try again prior to using them in their own classrooms is an important part of acquiring usable teaching skills with concrete materials and specific teaching approaches.

Another short-term project that can have enormously successful results for a department or school is a faculty retreat. The success that a particular curriculum enjoys can be traced largely to the articulation found in its delivery in the classroom. The chance for all the teachers in a school or a department to spend two to three days together discussing their goals, their methods, and their common successes is an important and worthwhile activity. It allows for the building of networks of communication within a school or department. In addition, it is an opportunity for teachers to reflect on the changing needs and conditions they face in their classrooms. Retreats offer the possibility of planning longer-term professional development programs that can be addressed specifically to identified needs.

Both the potential and the limitations of a short-term format need to be recognized. Short-term professional development activities are more effective if they are part of an ongoing cumulative program. Significant and lasting change usually requires a longer-term format.

LONGER-TERM FORMAT

Professional development programs requiring longer periods of commitment on the part of teachers are necessary to meet needs such as the following:

- Analyzing the mathematics curriculum for specified grades
- Planning a new course and preparing to implement it
- Developing an in-depth background in a given content area
- Meeting new certification requirements
- Implementing a new mathematics series
- Integrating technology across the mathematics curriculum
- Achieving articulation across, and within, a multisite, multilevel district

One common, and familiar, format for professional development is the course approach. This format closely resembles a university course on a specific topic. The method is quite appropriate for developing skills in a content area or reviewing or introducing pedagogical topics related to mathematics teaching. Some particular topics follow:

- Developing diagnostic skills (in whole number concepts, geometry, ratio and proportion, etc.)
- Strengthening the mathematics curriculum (in estimation, problem solving, using the hand calculator, applying decimals, etc.)
- Using technology in the mathematics classroom (hand calculators or microcomputers as concept developers, symbolic manipulation systems in algebra, etc.)
- Integrating such topics as discrete mathematics and algorithmics into the secondary school mathematics curriculum
- Illustrating applications of school mathematics in real-world settings

- Developing problem solving in the school mathematics curriculum—nature, related teaching skills, and evaluation

These are but a few illustrations of the wide variety of topics that could easily serve as a base for an entire course for a semester or longer. Traditionally, such courses have been delivered as extension courses by a local college or university, meeting one night a week for two or three hours. However, that is not the only way. Districts might think of partitioning these courses into six-week modules and offering them on flexible or floating schedules. For example, the course on problem solving might be split into modules dealing with problem-solving strategies and their application, teaching problem solving at a given range of grade levels, and evaluation of problem solving. In like manner, the course on applications might be split into applications of whole number arithmetic, applications involving real numbers, applications of geometry, or applications of other areas. Thus, teachers who have adequate knowledge of one or more modules of the content and prefer not to take the whole course would be motivated to take part(s) of it.

Longer-term formats are essential when teachers are expected to make major changes in the way they function in the classroom. Examples of such changes follow:

- Using manipulative-based instruction
- Making effective use of every minute
- Introducing cooperative learning
- Raising achievement through teacher expectations
- Using a problem-solving approach
- Teaching to objectives
- Using computerized instructional management
- Using computers and calculators

Major changes in teaching can be facilitated by a longer-term format that includes an input phase followed by an implementation phase. The input phase can consist of a course, a workshop, or a series of workshops. The implementation phase can consist of the following types of activities:

- A series of seminars
- Periodic meetings of a team of teachers
- Informal networking of participants
- Individual consulting by a workshop leader
- Classroom visitations by a resource person or peer

Including an implementation phase will help ensure that the desired changes actually take place in the classroom.

SMALL-GROUP FORMAT

The small-group format can lead to effective staff growth and cohesiveness. This form of professional development can employ a variety of approaches but requires a dedicated leader and a group that has a task focus. Small groups of individuals working together can jointly participate in such activities as the following:

- Developing curriculum
- Reading and discussing NCTM journal articles or positions
- Evaluating learning materials
- Making visits to observe exemplary teaching or programs
- Engaging in role playing or simulating classroom activities
- Comparing the relative efficacy of differing curricular or instructional approaches
- Discussing case studies of individual students and planning different ways of meeting students' needs
- Reading NCTM yearbooks or working through a new mathematics text or content area

Small groups with a specific goal, such as working to improve mathematics achievement in their school, lend themselves to the "quality circles" format. Such groups might be encouraged to work together to develop and implement a school-level plan for improving mathematics instruction. Successful ideas could then be used in the same district for staffs from other schools. These activities need to continue on a long-term basis, building a commitment to professional growth. New teachers need to be brought into the circle regularly to ensure that the movements of growth and leadership continue to spread throughout the staff and district.

Many activities can be designed to deal with small groups of teachers interested in growing professionally. One excellent opportunity for professional growth can come through involving teachers with curriculum specialists, university faculty, or others in carrying out a classroom-level research project on a question of interest to the school or district. Such activity tends to result in professional growth in student observation, structure of the curriculum, or knowledge of student patterns of learning mathematics. In addition to the knowledge gained, teachers gain an understanding about themselves that they can apply in the classroom.

The larger-scale reading group is also appropriate for work with small groups. Many collegiate mathematics departments use this model for staff growth in new areas, but the idea is not limited to the university level. Having faculty read together and discuss the material in light of their needs and conditions can do much to improve a given school program.

Coaching is a process that shows promise in the implementation phase of ongoing professional development activities. Coaching is a process in which professionals assist each other in improving instruction through the process of observation and feedback. Teachers should coach each other. In order to do this successfully, they need familiarity with the new skill or strategy, access to other teachers in their classrooms, and openness to suggestions. A process like coaching that involves the actual teaching while it is happening has a high probability of improving instruction.

INDIVIDUAL FORMAT

Many of the small-group activities and longer-term professional development plans can also be adapted to deal with an individual's plan for professional growth. Programs such as the following can foster individual growth in the ability to plan and provide quality instruction:

- A plan of program visitation
- Curriculum development through a sabbatical
- A summer internship in industry
- A temporary transfer to go to another school or another level within the district
- A personal reading program

These approaches call for dedicated self-motivators and trusting administrations. Strong leadership or group task orientation is needed for individual and small-group work to be successful.

These modifications in the traditional delivery system for courses must be considered to acquire teacher commitment and short-term continuity of the topics of discussion. These two factors are critical features essential for the success of any given professional development program.

PRACTICAL CONSIDERATIONS

The many and varied responsibilities carried by members of a school mathematics staff make long-term commitment to professional development activities hard to obtain. Flexible and innovative approaches to scheduling, packaging, and delivery are important features in getting the participation desired. Arrangements should be structured in such a fashion that they minimize the loss of a teacher's private or professional time. The effectiveness of after-school sessions tends to diminish when the sessions exceed two hours. Especially in sessions that are longer than one hour, some time should be given to active participation. The location should also be selected to minimize travel and "down time."

Longer activities should address the same features mentioned for shorter ones:

- Relevance to the teacher's perceived needs at the classroom level
- Clear goals for growth in knowledge and skills
- Relationships among theory, research findings, content, and teaching level should be highlighted and discussed.

Regardless of the format of the professional development, the desired changes have a much greater chance of being fully implemented in a school with a positive climate for change. Teachers need to have time to work together. Mathematics teachers should be assigned classes in close proximity to each other. The principal sets the tone by establishing the schedule, by recognizing and rewarding professional activities, and by creating an environment where teachers are encouraged to work together. Leadership by the department chair or other staff members is also important for establishing a climate in which teachers work together as professionals for the improvement of instruction.

SUMMARY

The foregoing suggestions provide a litany of alternative forms of professional development activities for meeting the needs of teachers and needs in the curriculum. The degree to which educational agencies, supervisors, and teachers focus their efforts toward meeting individual teachers' needs through professional development in the coming years will do much to define the success they have in maintaining a contemporary curriculum and successfully serving their students.

IMPLEMENTING THE PROGRAM

Professional development programs for teachers of mathematics should be effectively conducted and should include the following features:

a. A blending of mathematical content and effective pedagogy

b. Active participation of teachers

c. Attention to the concrete, day-to-day problems of teachers

d. An integration of theory and practical applications

e. Communication of objectives to participants

f. Opportunities for teachers to practice new skills and techniques in the classroom

g. Incorporation of support and follow-up activities

The most exciting part of a professional development program is the implementation phase, when the planning takes form and becomes evident in the actual event. This is the time when the research, needs assessment, teacher input, goals and objectives, and specific planning all come together. Regardless of the type of program offered, it is necessary to address the content, the audience or participants, and the mechanics in both the planning and implementation stages.

PARTICIPANTS

The National Staff Development Council includes the following questions in a checklist from its workshop "Conducting Effective Staff Development Programs." They serve as reminders to those who have responsibility for the various aspects of the in-service program to consider the audience or participants.

In implementing the in-service program, do you allow for—

- participants to make the role change from teacher to learner;
- participants to know "Who is here?" "Why are we here?" and "What resources are there?";
- a check with participants during the program concerning their attitudes, feelings, and knowledge;
- deviation from the plan to allow for meeting the emerging needs of participants;
- participants to work with a variety of people and other resources;
- participants to relate program outcomes to their own classroom settings?

Note that emphasis is placed on the critical aspect of the role change from teacher to learner. Being aware of this transition and the inevitable resistance of one or two staff members will help the in-service leader prepare for it with appropriate activities and responses. For example, the teacher who openly criticizes a new idea as unworkable might benefit from hearing a fellow teacher describe a successful experience with the idea.

MECHANICS OF THE WELL-PLANNED PROGRAM

Once the needs assessment provides information about the type of required program areas related to the designated content and to the specific groups of participants, the mechanics must be addressed. It's the small details that can make or break an otherwise well-planned program. Remember that not all the participants are as enthusiastic about the program as you are. Some of them might not even want to attend. For instance, don't give them an opportunity to reject the entire session just because the meeting room was too hot or too cold.

Here are some suggestions for the mechanics of implementating the program:

- Greet speakers and provide a quiet place for them to assemble and relax before the program begins.
- Inform the speakers in advance of any special scheduling, length of session, size of room, evaluation process, and plans for breaks.
- Monitor the sessions of the program during the presentations so that any unexpected incidents or exigencies can be dealt with efficiently.
- Clarify in writing the details of the recommendations and the plan for reimbursing the speakers' expenses. An agreement signed in advance will eliminate misunderstandings and create goodwill.
- Take advantage of the resources in your community and in business and industry to bring current applications of mathematics to the attention of the teachers.
- Make sure that materials used in the workshops are subsequently available for use in the classroom or that teachers know how to obtain them. This enables the teachers to replicate in their own classes what was learned in the sessions.
- Arrange for the exhibition of publishers' materials where appropriate.
- Consider the staff's comments and suggestions from previous in-service evaluations and the current sessions. Interviews or telephone calls to selected participants will yield important ideas as well as give the participants another opportunity to influence the content and details of the program. These contacts also make the teachers feel that their input as well as their participation is important.
- Verify the responsibilities of the participants: if they are expected to bring materials, send reminders.
- Provide the building principals with as much information about the programs as possible so that they, as instructional leaders, are able to be supportive advocates.
- Check the sound system and its effectiveness for the size of the audience before the session begins. Provide the best microphone for the situation so that all participants can hear and the speaker can be as mobile as desired.
- Assign someone to be responsible for the room temperature, telling him or her how to adjust the controls or whom to contact.
- Provide maps and post signs with clear directions to buildings and rooms, and indicate parking places. Make certain there is adequate parking space.

- Check the working order of equipment that is to be used by the leaders. Make sure there are extra light bulbs, pens, disks, pencils, and so forth.
- See that adequate outlets and tables at appropriate heights are provided for computers.
- Arrange computer demonstration setups for ease of viewing by participants.
- Make necessary security and insurance arrangements for computers and equipment.
- Include breaks at reasonable intervals for the comfort of the participants. Stretch breaks keep the blood flowing.
- Make announcements, post signs, or give maps for finding rest rooms, drinking fountains, and telephones.
- Provide facilities for hanging or storing wraps.

A checklist can ensure that the many tasks are assigned and completed on schedule. A sample checklist for planning meetings[2] is shown in Appendix 4. When a staff development activity is planned by a committee, a checklist is a guarantee that nothing is overlooked.

The checklist in Appendix 5 is for the presenter to use in planning the program[2]. If you have not previously tried such a list, do so with your next set of speakers and ask for their reaction to it. It is an excellent planning tool for the first-time speaker or the seasoned veteran. The program chair might supply some of the information if the speaker is unfamiliar with the target audience.

SPECIAL PREPARATIONS FOR SPECIFIC WORKSHOPS

Special-topic workshops will need special preparations. The following examples will give you an idea of what preparations might be needed.

Workshops Involving Solving Problems

Leaders of workshops that involve problem solving should consider the following suggestions:

- Model the teaching of problem solving.
- Teach problem-solving heuristics—drawing a picture, making a table, using trial and error, and so forth.
- Work the same problem several different ways, using different heuristics
- Obtain an abundant supply of interesting problems.

---

2. The checklists for planning meetings and presentations were developed by the Minneapolis Public Schools' Mathematics Department.

- Have a variety of problems that are suited to the various mathematical backgrounds of the participants.
- Create a friendly, nonthreatening atmosphere.
- Have each participant work on problems that are individually challenging but not overwhelming.
- Allot a large portion of the time for actually solving problems.
- Give participants opportunities to work cooperatively in small groups.
- Give participants ample opportunity to talk about solving problems.
- Emphasize estimation and reasonableness of results.
- Use calculators
- Have participants generalize results.
- Encourage participants to create problems.
- Give participants a variety of nontraditional problems to take with them.
- Give participants a list of sources of good problems.

Workshops Involving Equity Issues

Leaders of workshops designed to promote increased achievement and participation in mathematics by females or minorities should consider the following suggestions:

- Involve females and minorities as role-model leaders.
- Encourage participants to assess mathematics enrollment in their schools by sex and race.
- Encourage participants to assess the attitudes of their students toward mathematics.
- Provide a nonthreatening atmosphere.
- Model working together.
- Stress cooperative work in groups of two to six.
- Structure activities so that everyone is involved.
- Vary the activities to provide successful experiences to participants with varying strengths.
- Provide a setting where participants can be successful at solving problems.
- Stress high expectations.
- Provide information about mathematics and careers.
- Encourage participants to use equity materials distributed by NCTM.

Workshops Using Manipulatives

Professional development leaders who use manipulatives should consider the following suggestions:

- Provide research results on the use of manipulatives.
- Stress manipulatives as a means of developing concepts, not an end.
- Stress the use of spoken language as a bridge from the concrete to the abstract.
- Provide time for free exploration with manipulatives.
- Introduce one idea at a time. Allow time for participants to try it and then have them get back together for discussion.
- Have teachers try ideas in their classrooms between sessions.
- Stress the movement from concrete activities to visual activities to paper-and-pencil or mental activities.
- Involve leaders who have classroom experience with manipulatives.
- Arrange opportunities for teachers to work together by grade level; then they can pursue grade-level objectives in depth.
- Arrange times for teachers to exchange experiences.
- Provide information about classroom management of manipulatives.
- Provide "how to" materials.
- Introduce one manipulative at a time, allowing teachers time to get used to one before starting to use another.
- Provide opportunities for teachers to observe manipulatives in use.
- Help teachers obtain manipulatives for their classrooms.
- Provide for follow-up activities—seminars, classroom visitations, networking.

Workshops Using Computers

Leaders of professional development programs involving computers should consider the following suggestions:

- Whenever possible, arrange at least two sessions.
- Schedule sessions at least a week apart.
- Provide activities for participants to complete between sessions.
- Test the hardware and software before each session.
- Provide an agenda that includes topics, activities, times, breaks.
- Assess the prior knowledge and experience of the participants.
- Be alert to participants with computer anxiety.
- Draw on the strengths of participants.
- Plan for active involvement within the first half hour.
- Vary the modes—lecture, demonstration, discussion, hands-on activities.
- Provide many examples to illustrate significant concepts.
- Be flexible, adapting to feedback from participants.

- Relate to, and draw on, examples provided by participants.
- Stimulate active involvement—listening, doing, discussing, planning, revising.
- Model the desired behavior of a learner, and then have the participants model that behavior.
- During hands-on time talk to *every* participant individually.
- Schedule at most two participants to a computer.
- Stress mathematics, with the computer in a supporting role.
- Encourage experienced participants to help those with less experience.
- Provide reduced copies of transparencies for notes.
- Provide two copies of activity pages—one to use in the workshop and one to take home.
- Provide specific ideas for follow-up activities by the participants.

MAKING A PRESENTATION

The guidelines for presenters at NCTM meetings, listed in Appendix 6, have been developed from experience with many meetings over a long period of time. They contain practical ideas that are valuable for any presentation.

In addition, the following guidelines for use of visuals at NCTM meetings are useful for any presentation:

- Prepare visuals in advance.
- Use large lettering.
- Letter with a broad-tip marking pen or bulletin typewriter.
- Use short messages (about 15 words maximum).
- Use many visuals.

A presenter should determine from the sponsors exactly what is expected. The presenter may also wish to inquire about arrangements concerning equipment, room size, and setup. Communications about expectations and participants are probably more effective in person or by phone, whereas those about remuneration, expenses, and necessary facilities and equipment should be confirmed in writing.

The presenter should put the objective(s) of the presentation in writing. That tends to clarify the presenter's own thinking about what is to be accomplished and what new ideas, information, skills, or attitudes participants will gain as a result of this session.

The presenter should seek advance information about the participants and their level of prior knowledge and adjust the presentation to meet their needs. Some of this information can be gathered through informal discussions at registration time. The presenter can also do some assessing of the participants early in the session. This must be done in a manner that does not embarrass anyone. The same presentation that goes very well with one group can fall flat with another, because the prior knowledge and needs of the participants are different.

Presenters at professional development sessions should model effective teaching practices. For example, to involve the participants the leader can divide them into groups of two to six and model a cooperative learning experience. The involvement needs to be carefully structured so that the activities are meaningful and "on task." Logistics of hands-on activities must proceed smoothly. One effective practice is to involve the participants. Plans for involvement need to be especially well thought out. The logistics of hands-on activities must proceed smoothly. This involvement in discussion needs to be structured so that the discussion is meaningful and "on task."

Often there is much more information to be presented than time allows. When this happens, participants can be given portions of the information as examples and be motivated to continue to seek more information. When time is short, impressions are especially important; they tend to be more lasting than factual information.

Presenters need to have a plan for pacing the presentation so it will fit the allocated time. A presentation needs an opening to introduce the topic and establish rapport with the audience. It also needs an ending that summarizes the essential ideas and finishes on a positive note. The body of the presentation that comes between should be planned flexibly so that material can be added or deleted as time allows. First-time presenters tend to be afraid of not having enough material to fill the time, so they overprepare. Then as a result of the runaway involvement of participants or other factors they can find themselves running out of time. When this happens, the common error is to try to present the bulk of the material in the last five minutes, thereby inundating the participants with much more than they can assimilate. A better procedure is to delete some of the material and inform the audience of sources for learning more about the topic.

Presenters should have self-evaluation plans for each session they present. One idea is to use the results from an evaluation by the sponsor. Another idea is to use one's own simple self-evaluation form. Yet another is to ask someone to participate in the session and then get together with you afterward to discuss it. Through continual evaluative feedback, presenters can develop into polished performers.

To remind presenters of the ideas enumerated above, a sample presentation planner appears in Appendix 5.

AFTER THE PROGRAM

Immediately after the in-service program someone needs to—

- straighten up the meeting space;
- remove used coffee cups and other debris;
- clean and return the coffee urns and serving plates;
- return equipment, extra handouts, and supplies;
- collect evaluations and attendance information.

These details can be accomplished quickly if responsibilities are assigned in advance and everyone involved in the planning pitches in to help.

The coordinator should see that the following business details are processed promptly:

- Reimbursements
- Attendance reports
- Evaluation reports
- College credit (if applicable)
- Thank-yous
- Other information generated at the program

Thank-yous are especially important to ongoing professional development. Individuals who feel that their contributions are appreciated are more likely to contribute when called again. Individual thank-yous are more effective than form letters. The principal and the school district personnel department should receive copies of letters to teachers. The positive reinforcement that teachers receive contributes to their motivation for professional development. Presenters should be provided with the evaluation information on their sessions.

Shortly after the meeting, the planners should get together to go over the evaluations and discuss results. The ideas generated at these sessions contribute to improving future professional development events.

Detailed planning pays off. A well-planned program runs smoothly because the participants are able to focus on the content and not on minuscule problems that would tend to interrupt or divert their attention. Participants in a well-implemented program usually take the mechanics for granted, but the program coordinator knows that they are all part of the plan.

EVALUATING THE PROGRAM

Professional development programs for teachers of mathematics should be systematically evaluated, with attention to these issues:

a. Determining whether the needs they are designed to meet have been satisfied

b. Using the results from the evaluation to improve and develop future programs

These issues from NCTM's position statement on professional development are generally supported by anyone who is planning professional development activities for teachers. The challenge is to put both the letter and the spirit of these issues into practice.

Arranging a trip, playing football, teaching a mathematics lesson, or evaluating a professional development effort all require a plan. A well-designed plan of action can lead to a successful vacation, a good football game, an effective mathematics lesson, or a meaningful professional development program evaluation.

EVALUATION PLAN

Evaluation should be planned well in advance of a professional development program and should not come as a surprise to presenters, participants, or program decision makers. An evaluation plan should include the following:

- Evaluation objectives
- Evaluation design
- Data analysis and interpretation
- Writing and distributing the report(s)

OBJECTIVES

When the objectives of the evaluation are being planned, the two primary purposes for evaluation, as stated in the NCTM position, should be considered. Each will generate specific reasons for the evaluation, as shown in the examples below:

a) Determining whether the needs they are designed to meet have been satisfied:

Did the activity meet its objectives?

How effective was the process used?

Was the use of resources justified?

Should the use of resources be continued?

b) Using the results from the evaluation to improve and develop future programs:

How can we provide feedback to participants?

How can we provide feedback to presenters?

How can we communicate with clients and decision makers?

How can the evaluation results help improve future programs?

The objectives of the evaluation are determined by both the objectives of the professional development activity and the reasons for evaluating the activity. The purposes of the evaluation will influence both the initial objectives that are set and the final reports or documents that are developed.For example, if an objective of the professional development activity isto increase the number of different instructional strategies used by mathematics teachers, then an objective of the evaluation could be to determine whether teachers increase the number and frequency of the different instructional strategies they use after participating in the activity.

If the activity made reference to *Arithmetic Teacher* articles concerning different instructional strategies, the evaluation could include checking with the librarian to see how often copies of the *Arithmetic Teacher* were checked out by teachers after the activity.

If the objective is to increase the effectiveness of teaching geometric and measurement concepts in the elementary classroom, then an objective of the evaluation could be to increase the percentage of students who master the geometry and measurement objectives on the California Test of Basic Skills (CTBS) Objectives Mastery Report. The evaluation could then include a comparison of the percentage of students who mastered the geometry and measurement objectives on the Objectives Mastery Report for classes before and after teachers participated in the activity.

Evaluation objectives can be written for any professional development activity. Such objectives help to determine the effectiveness of the effort. Once the objectives have been identified, then the design of the evaluation itself can be undertaken.

DESIGN

The evaluation design may include one or all of the following components or stages:

- Preassessment
- In-process data collection
- Postassessment
- Follow-up

Preassessment

A pretest, or preassessment, may be helpful for collecting needs assessment information that can be used both to structure the activity and to evaluate it. For example, an evaluation design for the objective "to increase the number of instructional strategies used by teachers" might include the preassessment activity shown in figure 2.

The information obtained from this assessment can be used to structure the professional development activity itself. The responses will show how the participants define "instructional strategy" and will also indicate the kinds of instructional strategies they presently use and how often.

Please list the instructional strategies you now use in your mathematics class-room.

For each strategy, please indicate what percent of your mathematics instructional time is spent using that particular strategy.

Instructional Strategy	% of Time Used
1. Situational lessons	1. 20%
2. Guided practice	2. 10%
3.	3.
4.	4.
5.	5.
•	•
•	•
•	•

Fig. 2

In-Process Data Collection

In-process data collection can be used to obtain feedback from the participants concerning the effectiveness of the activity while it is being conducted. This feedback can be structured or unstructured. Here is an example of a somewhat unstructured approach:

> The leaders give each participant a blank 5 × 7 card with these instructions: "At the end of our session today, please take time to write comments to us concerning what you liked or disliked about today's session, changes that should be made in the workshop, and any other comments that would be helpful."

In a structured approach, specific questions can be listed on a form. A sample form appears in Appendix 7.

Postassessment

The postassessment is conducted some period of time after the professional development activity. It should solicit information about the participants' perceptions of the effectiveness of the activity as well as data that could be used to refine the activity itself the next time it is conducted. A sample postassessment instrument is shown in figure 3. Figure 4 shows an example of a postactivity experience survey.

Follow-up Evaluation

Follow-up evaluation is generally conducted several weeks after the activity. Interviews, observation, and survey instruments can be used to collect

POSTASSESSMENT

1. Please list the instructional strategies you now plan to use and the amount of time you plan to allot for each.

    Instructional Strategy   *% of Time*

    _____   _____

    _____   _____

    _____   _____

Answer questions 2–7 on a scale of 1 to 5.

1. No, or the most negative response
5. Yes, or the most positive response

                  NO   YES

2. Did this activity increase the number of instructional strategies you will use in your mathematics classroom?   1   2   3   4   5
3. Were the objectives for this activity clear?   1   2   3   4   5
4. Did the activities match the objectives?   1   2   3   4   5
5. Were the presenters prepared?   1   2   3   4   5
6. Were the presenters organized?   1   2   3   4   5
7. How would you rate this activity?   1   2   3   4   5
8. What did you like most about this activity?

9. What did you like least about this activity?

10. How would you improve these sessions?

Additional comments: _____

Fig. 3

follow-up information. For example, the teachers who participated might be interviewed as follows:

POSTACTIVITY EXPERIENCE SURVEY

1. How useful was the in-service activity to you?

    Not useful   Very Useful

    1   2   3   4   5

2. List the specific strategies you have implemented as a result of the in-service activity.

    _____

    _____

    _____

3. When did you implement these strategies?

    After

    Two Months   Immediately

    1   2   3   4   5

4. Overall, how effective were the activities you have implemented?

    Don't Know Minimally Very

    1   2   3   4   5

5. List the strategies you plan to implement in the future.

    _____

    _____

    _____

6. Did you receive follow-up assistance (visit, phone call, peer support, etc.)? _____

    _____

    If yes, how useful was the assistance?

    Not Useful   Very Useful

    1   2   3   4   5

7. Comments: _____

    _____

    _____

Fig. 4

- Are you using any of the ideas or strategies you learned in the workshop?
- If yes, which strategies, and when do you use them?
- If no, why not?
- How would you rate the long-term usefulness of the workshop?

If time permits and staff is available, on-site observation is an excellent way to determine whether the ideas or strategies that were presented are being used by teachers in their classrooms. The observation also gives a staff developer or leader an opportunity to coach and encourage teachers as they attempt to make the prescribed changes.

Newsletters and telephone calls are effective means of making follow-up contact with participants. Evaluation data, motivational articles, ideas that participants have found successful, and other related news can be featured.

A survey of participants' postactivity experiences provides useful evaluation information. It may also guarantee that ideas presented in the professional development activity will be used in the classroom if the participants are aware that such a survey will be taken. When interoffice mail cannot be used to distribute and collect the surveys, each person can be asked to self-address an envelope. Several weeks after the workshop, leaders can mail the survey to participants. To encourage a good response, a return envelope addressed to the activity leader might be enclosed.

For maximum effectiveness and ease in compiling the results, evaluation instruments should be succinct and easy to complete. Rated responses meet these specifications. Rating scales can include numbers and words, as in the example below. Consistency should be maintained in the arrangement of negative to positive items.

vague — 1 — 2 — 3 — 4 — 5 — clear

very poor — 1 — 2 — 3 — 4 — 5 — very good

never — 1 — 2 — 3 — 4 — 5 — always

always — 1 — 2 — 3 — 4 — 5 — never

chaotic — 1 — 2 — 3 — 4 — 5 — clear

slow — 1 — 2 — 3 — 4 — 5 — fast

abstract — 1 — 2 — 3 — 4 — 5 — concrete

boring — 1 — 2 — 3 — 4 — 5 — challenging

DATA ANALYSIS AND INTERPRETATION

Data can be tabulated and presented on the original evaluation form. This is particularly true for inventories, questionnaires, and checklists. Such data can also be presented visually, with graphs or charts. This is actually made quite easy with access to personal computers. Following is a sample of evaluative information that was generated by a personal computer. Figure 5 displays the mathematics scores on a standardized test for the class of 1989 in grade-level equivalents. Figure 6 shows the teacher responses to a postassessment survey of a professional development activity.

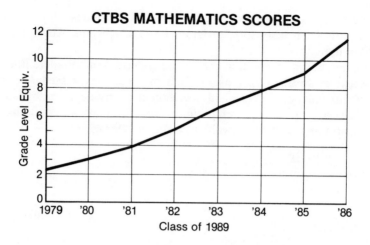

Fig. 5. CTBS scores in grade-level equivalents

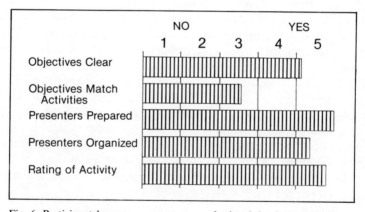

Fig. 6. Participants' average response to professional development program

Responses from interviews and open-ended questions are more difficult, but not impossible, to tabulate. Since the data do not fall into categories, it may be effective to present them in a descriptive form. Personal reactions or comments made by participants may also be presented in this way.

It is effective to include anecdotal evidence as well as statistical support in your evaluation. The anecdotal reports about particular activities or participant behavior can add interest, as well as credence, to the interpretations.

WRITING THE REPORT

The purpose of the evaluation report is to provide information to persons who make decisions about continuing or changing the professional develoment program. The target audience for the report includes administrators, boards of education, sponsors, program leaders, teachers, and interested community members.

It is important that evaluation reports be concise, but complete information is necessary for them to be useful. In light of often-expressed skepticism of the effectiveness of professional development activities, written reports are valuable documents for summarizing accomplishments and building support for continuing activities. Evaluation reports should include—

- program objectives;
- program description;
- evaluation procedures;
- results;
- conclusions.

This information should be presented with the reader in mind and in a manner that is fair and accurate. Any educational language, or jargon, should be translated into commonly used language. If the report is written for teachers, they will want to use it to make instructional decisions. They will be interested in the major successful and unsuccessful activities, any unanticipated events that may have influenced results, and the program conclusions.

Administrators are decision makers who will certainly have influence, if not control, over continued or increased funding. They need to understand the evaluation plan in order to respond to questions. A summary of the report, including the findings and conclusions, would be a valuable resource for administrators. Superintendents, boards of education, and interested community members are also potential readers of an evaluation report. They will be interested in both statistical support for conclusions and the anecdotal evidence for them.

CONCLUSION

Evaluation is an important aspect of a professional development activity and should be included in the initial planning. The purpose, design, and use of the evaluation should be developed and shared with presenters, participants, and decision makers in mind. The intent of the evaluation should be to improve and refine the activity so that the information learned can be generalized to make future activities more effective.

There is an analogy between solving a problem and conducting a professional development program. Consider Polya's four phases of good problem solving:

1. Understanding the problem
2. Devising a plan

3. Carrying out the plan
4. Looking back at the solution

Making an assessment and stating the goals and objectives corresponds to understanding the problem. Planning a professional development program and selecting a format correspond to devising a plan for a problem. Implementing the program corresponds to carrying out the plan for the problem. Finally, evaluating the professional development program corresponds to looking back at the solution of the problem. In view of this analogy, mathematics specialists should be successful when they apply their problem-solving skills to planning, implementing, and evaluating professional development programs.

PART III
FINDING RESOURCES

RESOURCES

Numerous resources are available to help mathematics teachers, supervisors, and administrators develop professionally. These resources come in many forms and from many sources.

Local School Districts

Many school systems provide extensive professional development opportunities for their employees, such as the following:

- *In-service courses, workshops, seminars, and other information-sharing and competence-developing activities.* Many such activities are in response to special requests by teachers.
- *Professional or staff development centers containing up-to-date professional books, journals, and nonprint materials, such as films, filmstrips, videotapes, software, and records, that focus on enhancing a teacher's knowledge.*
- *Local supervisors and other personnel who offer advice on specific classroom problems faced by the teacher.* Supervisors may also demonstrate new materials and methods on request.
- *Career and crisis counseling services.* These services seek to help teachers understand the career they have chosen and to relate realistically to its demands.

Colleges and Universities

Colleges and universities play a major role in defining the professional status of teachers through preservice programs and efforts to help practicing teachers maintain their professional standards. Among these services are the following:

- *Access to professional libraries and faculty who are interested in assisting with problems and opportunities related to mathematics teaching.* Professors are often involved in research directly related to professional development.
- *Content and professional education courses for mathematics teachers.* Specialized teaching programs, often leading to advanced degrees, are frequently available.

- *Specialized courses, workshops, and seminars designed for a specific group of teachers and taught in their schools.* Specialized conferences that provide opportunities for mathematics teachers to learn about current research in teaching and learning and to gain insight into innovative technologies for teaching are also held on campus.
- *Sponsorship of local and state mathematics contests.* Such activities provide an opportunity to involve many students.

State Departments of Education

State departments of education provide a myriad of services to school districts, and thus to mathematics teachers. As state agencies, they must be responsible for (and responsive to) the needs of every local school district in the state. Hence, it is their job to provide information about certification requirements and regulations and put you in contact with other persons who can respond to your requests. Usually, state departments of education provide some of these services:

- Technical and professional assistance to local school systems in relation to testing, curriculum development, the identification of educational needs, and issues related to teacher certification, supply, and demand.
- Assistance in identifying, planning, and implementing projects within local school systems.
- Coordination of large-scale testing and evaluation, such as functional mathematics tests and high school equivalency tests.
- Service as a conduit through which federal and state monies for educational improvement flow and coordinator of other national programs supporting mathematics education goals.

Professional Organizations

There are many organizations that can help educationally oriented personnel develop professionally. A major goal of the National Council of Teachers of Mathematics is to assist mathematics teachers in their quest of becoming knowledgeable, highly skilled professionals. The professional staff of NCTM, located at 1906 Association Drive, Reston, Virginia 22091, is available to assist you on a wide range of educational concerns. When you become a member of NCTM, the following services are available to you:

- *Professional mathematics education publications*, including the *Arithmetic Teacher*, the *Mathematics Teacher*, the *Journal for Research in Mathematics Education*, the *NCTM Newsletter*, yearbooks, and a variety of other publications that focus on critical topics in mathematics teaching and learning
- *Conferences and conventions* at reduced registration rates. Members

are invited to share ideas and teaching methods through involvement in conference programs.

- *Opportunities* to become involved in mathematics education on a national level through committee membership, presentations at conferences, and publication in the journals, yearbooks, and other education materials of the NCTM

While NCTM serves the profession on a national and international level, its affiliated groups attempt to focus on the professional development of mathematics teachers on the local, state or provincial, and regional levels. Affiliated groups provide many services to their membership, including the following:

- Frequent meetings that provide opportunities for sharing ideas and keeping current on the professional interests of mathematics educators
- Regional and state mathematics education conferences that offer a variety of opportunities for professional growth
- A journal to which members may submit articles or descriptions of successful teaching activities
- Recognition for outstanding mathematics teachers
- Awareness of local and state issues and opportunities to implement action on a local and state scale
- A sense of professionalism among teachers in a local area, developed through mailings, telephone calls, and participation in the working of the affiliated group
- The opportunity to have an impact on the national scene through resolutions presented by the affiliate's representative to the NCTM Delegate Assembly

Special professional needs of school district mathematics supervisors are addressed by the National Council of Supervisors of Mathematics, which is affiliated with NCTM. State supervisors are served by another NCTM affiliate, the Association of State Supervisors of Mathematics. The Mathematical Association of America (MAA) provides services for mathematics teachers at the college level.

FUNDING AND OTHER SUPPORT

In addition to these resources for mathematics teachers, new and exciting sources of support, financial and otherwise, can be identified, many from the private sector. Characteristic of many of these is their support of curriculum improvement projects, which usually involve the students, the teacher, principal, mathematics supervisor, and members of the community. In some instances, all of the previously mentioned groups (local school districts, colleges and universities, state and provincial departments of education, and professional organizations, such as NCTM and its affiliated groups) work together to plan and implement school-based projects. Some sources of support are listed below.

Business and Industry

Many businesses and industries have expressed strong interest in improving the state of mathematics education in elementary, middle, and secondary schools. This interest usually exhibits itself in various types of support, including the following:

- Resource persons, such as speakers, who can offer technical advice to teachers or develop topics for students
- Opportunities for teachers or students to become involved in business or industry as part-time employees or through cooperative educational programs
- Loans or grants for equipment for projects related to business and scientific interests
- Support for educational improvement efforts, such as teacher research projects, scholarships for further study, teacher recognition, or sponsoring a professional meeting
- Donation of equipment and materials

Private Foundations

Private foundations support many projects related to educational improvement. As a general rule, such foundations target funds for specific areas, such as the study of issues related to women and mathematics or minorities and mathematics. Many private foundations award scholarships to support teachers' participation in special institutes or study abroad.

Federal Agencies

Federal agencies tend to focus on national concerns and to support projects and developments that will have a national or regional impact. The following list suggests ways in which you can become involved in activities of such federal agencies as the National Science Foundation or the Department of Education.

- Submit your name as a possible reviewer of proposals that address mathematics education issues. This is a firsthand way to become aware of current issues and promising approaches to teaching.
- Request guidelines for submitting proposals and information about eligible areas of funding or projects already funded in which you are eligible to participate.

- Request copies of final reports of completed projects that relate to your concerns. This is a good way to keep up to date with research in your field.
- Join forces with other teachers, supervisors, or local college personnel and submit a proposal for funding.

Community Resources

Every community has resources that can assist you in your professional growth. Possible community resources might include the following:

- A wealth of persons interested in mathematics education. This list would include parents, officers in mathematics-oriented groups, teachers, retired scientists or engineers, and professionals from banking, insurance, or industry.
- Teachers from community colleges, four-year colleges, and universities.
- Organizations, such as sororities and fraternities, that have an interest in educational or social concerns. Civic groups such as Rotary, Lions, and women's clubs also have educational interests.
- Physical resources such as museums, high-tech industries, harbors, communications, installations of the Federal Aviation Administration, federal laboratories, military installations, and utility companies offer good scientific information through tours, printed material, and other presentations.

Publishers

Commercially published instructional materials, primarily textbooks, represent a valuable resource for mathematics teachers. Teachers rely on the textbook for an orderly presentation of generally acceptable mathematics content and on teachers' editions for supplementary content and teaching strategies. However, it isn't always clear how teachers can influence the content of textbooks or the way the content is presented. As a general rule, publishers welcome ideas that have the potential for improving mathematics education. Many publishers periodically survey teachers and supervisors, and their local representatives request reports of district and state textbook adoption evaluations. Such evaluation activities offer an excellent opportunity for professional growth.

Most publishers provide workshops or specially prepared packages to acquaint teachers with the philosophy, content, teaching strategies, and special features of their products. Either of these can be the basis of a self-study or group professional development activity.

STRATEGIES FOR SEEKING RESOURCES

- In seeking resources, use the FOWGI principle—Find Out Who's Got

It! Make yourself aware of the various sources of funding and other resources that are available. Larger school districts have staff members who monitor funding sources by following the Federal Register and other sources of information.

- Place a priority on local resources. The closer the resources are to you, the more influence you can have on them. Local resources are less likely than remote resources to dry up for reasons completely beyond your control.

- Look for resources that will continue to be available year after year if needed. Local school district funds tend to be available on an ongoing basis except for times of budget crisis (which occur frequently in some districts). U.S. Department of Education Chapter 1 (formerly Title 1) funds to serve disadvantaged students have been available for the last twenty years; however, funding from other federal programs involved with mathematics education have tended to come and go.

- Distinguish between "hard money" and "soft money." Hard money is available year after year and is used to fund ongoing programs. Soft money tends to be available for only a limited period of time (one to five years) and is therefore appropriate for relatively short-term professional development activities. The benefits of these activities can continue to be felt in the schools long after the funding has ceased.

- Work for early success—particularly with short-term, year-to-year funding.

- Continued funding is easier to secure when the results of the professional development program are already having an impact in classrooms.

- Place a priority on activities that produce maximum results for minimum cost. Being able to demonstrate a cost-effective activity gives credibility to a request for additional resources.

- Locate organizations and individuals with objectives similar to your own. Work with them on activities whose objectives most closely coincide. School districts can collaborate and merge funds to produce stronger and more effective staff development programs. Colleges and universities with professional development courses have a special interest in working cooperatively with the schools whose students they receive.

- Proceed with the staff development activities in a manner that develops a feeling of "ownership" among the teachers participating. Testimonials from satisfied teachers make a strong case for continuing and increasing the funding.

- Conduct an assessment and use it to demonstrate the need for professional development activities. Assessment results can be publicized among both the professional staff and potential sources of external support.

- Make effective use of publicity. Don't promise in advance that a particular staff development project will solve all problems. You could be setting yourself up for a letdown. However, once success is achieved, let the world know about it. Also, arrange to recognize participating teachers in their local schools and their local press.

SUMMARY

To conduct a satisfactory professional development effort, you usually need imagination, determination, and money. With enough imagination and determination you will probably get the money. Many resources are available to help teachers develop professionally. Some are so obvious that they may be easily overlooked. The list of resources above by no means exhausts the possibilities; it is meant to give you some ideas to which you can add. Your charge: use ideas from this handbook for professional development activities that will inspire teachers and benefit students.

APPENDIXES

A Position Statement on. . .

PROFESSIONAL DEVELOPMENT PROGRAMS
FOR TEACHERS OF MATHEMATICS

Teachers of mathematics, like all professionals, require ongoing and cumulative professional development programs that enhance and maintain their teaching skills and knowledge. Because mathematics and education are disciplines that grow and change, teachers cannot depend on what they learned as undergraduates to carry them through their entire careers. Findings of research continually increase our understanding of teaching and learning. Further, social and technological changes increase the average citizen's need to understand and use mathematics. These forces demand reconsideration of the content and methods of mathematics instruction.

Curricular and instructional changes, however, do not occur automatically. The extent to which new ideas and techniques are integrated with current classroom practices depends on teachers' knowledge, motivation, and commitment to continued professional growth. The improvement of mathematics programs depends on well-prepared and well-informed teachers.

Such changes and improvement require teachers to have opportunities for high-quality professional development. The provision of these opportunities, which should maintain, enrich, and improve the skills and abilities that teachers need to serve their students best, is the shared responsibility of districts, schools, and individual teachers.

To help promote high-quality classroom instruction in mathematics, the National Council of Teachers of Mathematics encourages and supports the development and implementation of comprehensive professional development programs. The Council recommends that such programs be developed in accord with the following guidelines:

1. Professional development programs for teachers of mathematics should be based on a strong commitment to professional growth.

    a) An appropriate person should be responsible and accountable for the professional development of the teachers.

55

   b) Sufficient time should be allocated for individuals to assess needs, plan activities, lead or participate in programs, and evaluate outcomes.

   c) Sufficient funds should be available to support professional development programs and ensure teachers' participation in them.

2. Professional development programs for teachers of mathematics should be carefully planned.

   a) Clear objectives should be established.

   b) The programs should improve students' learning experiences by improving the skills and knowledge of their teachers.

   c) Those whom the programs are designed to assist should contribute significantly in planning the programs.

   d) Extensive assessments of individual and collective needs should serve as bases of the programs.

   e) Current concerns and issues in mathematics education should be reflected in the content of the programs.

   f) The programs should be ongoing and cumulative.

3. Professional development programs for teachers of mathematics should recognize individual differences.

   a) Varied formats, including workshops, conferences, institutes, courses, and in-school discussion sessions, should be used.

   b) Programs should be tailored to meet the needs of teachers whose knowledge, skills, and experiences are diverse.

4. Professional development programs for teachers of mathematics should be effectively conducted and should include the following features:

   a) A blending of mathematical content and effective pedagogy

   b) Active participation of teachers

   c) Attention to the concrete, day-to-day problems of teachers

   d) An integration of theory and practical applications

   e) Communication of objectives to participants

   f) Opportunities for teachers to practice new skills and techniques in the classroom

   g) Incorporation of support and follow-up activities

5. Professional development programs for teachers of mathematics should be systematically evaluated, with attention to these issues:

   a) Determining whether the needs they are designed to meet have been satisfied

   b) Using the results from the evaluation to improve and develop future programs

READINESS, PLANNING, TRAINING, IMPLEMENTATION, MAINTENANCE (RPTIM) MODEL PRACTICES

The RPTIM Model Practices offer a comprehensive research-based approach to designing and implementing effective professional development programs. It is predicated on the ten critical assumptions that follow and offers thirty-eight specific practices.

Basic Assumptions

1. All school personnel need in-service [education] throughout their careers.
2. Significant improvement in educational practice takes considerable time and long-term in-service programs.
3. In-service education should focus on improving the quality of school programs.
4. Educators are motivated to learn new things when they have some control over their learning and are free from threat.
5. Educators vary widely in their competencies and readiness to learn.
6. Professional growth requires commitment to new performance norms.
7. School climate influences the success of professional development.
8. The school is the most appropriate unit or target of change in education.
9. School districts have the primary responsibility for providing the resources for in-service training.
10. The principal is the key element for adoption and continued use of new practices and programs in a school.

RPTIM MODEL PRACTICES

Stage I: Readiness

1. A positive school climate is developed before other staff development efforts are attempted.
2. Goals for school improvement are written collaboratively by teachers, parents, building administrators, and central office administrators.
3. The school has a written list of goals for the improvement of school programs during the next three to five years.
4. The school staff adopts and supports goals for the improvement of school programs.
5. Current school practices are examined to determine which ones are

congruent with the school's goals for improvement before staff development activities are planned.

6. Current educational practices not yet found in the school are examined to determine which ones are congruent with the school's goals for improvement before staff development activities are planned.

7. The school staff identifies specific plans to achieve the school's goals for improvement.

8. Leadership and support during the initial stage of staff development activity are the responsibility of the principal and central office staff.

Stage II: Planning

9. Differences between desired and actual practices in the school are examined to identify the in-service needs of the staff.

10. Planning of staff development activities relies, in part, on information gathered directly from school staff members.

11. In-service planners use information about the learning styles of participants when planning staff development activities.

12. Staff development programs include objectives for in-service activities covering as much as five years.

13. The resources available for use in staff development are identified prior to planning in-service activities.

14. Staff development programs include plans for activities to be conducted during the following three to five years.

15. Specific objectives are written for staff development activities.

16. Staff development objectives include objectives for attitude development (new outlooks and feelings).

17. Staff development objectives include objectives for increased knowledge (new information and understanding).

18. Staff development objectives include objectives for skill development (new work behaviors).

19. Leadership during the planning of in-service programs is shared among teachers and administrators.

Stage III: Training

20. Staff development activities include the use of learning teams in which two to seven participants share and discuss learning experiences.

21. Individual school staff members choose objectives for their own professional learning.

22. Individual school staff members choose the staff development activities in which they participate.

23. Staff development activities include experiential activities in which participants try out new behaviors and techniques.
24. Peers help to teach one another by serving as in-service leaders.
25. School principals participate in staff development activities with their staffs.
26. Leaders of staff development activities are selected according to their expertise rather than their position.
27. As participants in staff development activities become increasingly competent, leadership behavior becomes less directive or task-oriented.
28. As participants in staff development activities become increasingly confident in their abilities, the leader transfers increasing responsibility to the participants.

Stage IV: Implementation

29. After participating in in-service activities, participants have access to support services to help implement new behaviors as part of their regular work.
30. School staff members who attempt to implement new learnings are recognized for their efforts.
31. The leaders of staff development activities visit the job setting, when needed, to help the in-service participants refine or review previous learning.
32. School staff members use peer supervision to assist one another in implementing new work behaviors.
33. Resources are allocated to support the implementation of new practices following staff development activities (funds to purchase new instructional materials, time for planning, and so forth).
34. The school principal actively supports efforts to implement changes in professional behavior.

Stage V: Maintenance

35. A systematic program of instructional supervision is used to monitor new work behavior.
36. School staff members utilize systematic techniques of self-monitoring to maintain new work behaviors.
37. Student feedback is used to monitor new practices.
38. Responsibility for the maintenance of new school practices is shared by both teachers and administrators.

SAMPLE CHECKLIST FOR SUPPLIES

_____ Handouts

_____ Prepared transparent visuals

_____ Easel and paper pad

_____ Markers

_____ Overhead projector/screen/extra bulbs/extension cords

_____ Markers for overhead projector

_____ Blank transparencies

_____ Microphone

_____ Sponge

_____ Paper towels

_____ Stapler

_____ Scissors

_____ Clipboard

_____ Sharpened pencils

_____ Copy-not pens

_____ Masking tape

_____ Paper

_____ Refreshments

_____ Coffee, tea, cream, sugar

_____ Disposable cups

_____ Napkins

_____ Trash receptables

SAMPLE CHECKLIST FOR PLANNING MEETINGS

Title of meeting _____

Participants _____

Date of meeting _____ Time _____

Day of meeting _____ Anticipated attendance ____

Place of meeting _____

Purpose of meeting _____

Estimated budget _____

Tasks	Person Responsible	Deadline Date	Task Date

Before Meeting

Reserve meeting space _____

Contact speaker(s) _____

Allocate time and space _____

Develop program description _____

Send out announcement(s) _____

Prepare printed agenda or program _____

Prepare handout materials _____

During Meeting

Registration (sign-in, name tags) _____

Attendance _____

Coffee, tea, refreshments, meals _____

AV equipment _____

Setup of room(s) _____

Chair meeting person _____

Host guest speaker(s) _____

After Meeting

Clean up _____

Return equipment and materials _____

Minutes, attendance reports _____

Reimbursements _____

Evaluation report _____

Thank-yous _____

SAMPLE PRESENTATION PLANNER

Title _____

Description _____

_____

Date _____ Time _____ Place _____

Room name/number _____ Room capacity _____

Room setup _____

Equipment _____

Visual aids _____

Handouts _____

Objective(s) of presentation _____

Impression(s) to be conveyed _____

_____

_____

Target audience _____

Anticipated level of prior knowledge _____

Expectations of meeting sponsor _____

_____

Plans for participant involvement (if any) _____

_____

Presentation plan:

    Opening _____

    Body _____

    Closing _____

Pacing:

    What to omit if running long _____

    What to use for "filler" if running short _____

Plan for self-evaluation of session _____

_____

_____

GUIDELINES FOR SPEAKERS

The NCTM is continually striving to maintain the high quality of its programs. This depends greatly on the success of those who give presentations at each conference. You may find the following guidelines helpful as you prepare your presentation.

1. Complete a "Biographical Data Form" and return to the NCTM office in Reston.
2. Be sure handouts are printed clearly and are in sufficient quantity.
3. If you decide to use an overhead projector, be sure your transparencies are readable in a large room. Bring along additional blank transparencies and pens.
4. Plan well in advance to ensure that needed materials are delivered to the meeting. We suggest that you carry them by hand if possible.
5. Plan to arrive at your assigned room early to check on your requested equipment.
6. Begin and end on time.
7. To maintain professional standards, it is inappropriate to promote commercial materials for personal gain during your presentation.
8. In keeping with NCTM policy, no disks are to be sold in computer workshops.
9. You are urged to avoid the use of language that can be construed as being sexist or degrading to a racial or ethnic group.
10. When you have finished your presentation, plan to leave quickly to allow the next presenter time to prepare.
11. In the remote possibility that you find it necessary to cancel your presentation, please notify the person chairing the program immediately by telephone.

Thank you in advance for your willingness to contribute your time, energy, and expertise to the enrichment of mathematics education. We trust your experience will be a positive one.

WORKSHOP/COURSE EVALUATION

Workshop or course title: _____

Please rate this workshop or course in terms of the areas listed below using the following scale:

> 5: Excellent—outstanding
> 4: Good—better than average
> 3: Satisfactory—okay, average
> 2: Fair—below average
> 1: Poor—markedly inadequate

Circle *ONE* number for each item:

1. Opportunities for learning new ideas, methods, or specific skills:
       5   4   3   2   1

2. Applicability to your own educational setting. . .classroom, school:
       5   4   3   2   1

3. Motivation to use what you have learned in your own situation:
       5   4   3   2   1

4. Stimulation to continue learning in the workshop or course content area:
       5   4   3   2   1

5. Opportunities for your own active involvement in the workshop course activities:
       5   4   3   2   1

6. Organization. . .flow of workshop or course program:
       5   4   3   2   1

7. Quality of presentation:
       5   4   3   2   1

8. Adequacy of facilities:
       5   4   3   2   1

9. *Overall rating*:
       5   4   3   2   1

Comments: Any comments you may wish to make on the reverse side will be appreciated.

BIBLIOGRAPHY

Childs, Leigh. *San Diego Math Network Training Guide*. San Diego, Calif.: San Diego County Office of Education, 1985.

Connecticut State Department of Education. *Professional Development Planning Guide: A Primer for Social School Districts*. Hartford, Conn.: The Department, 1984.

National Council of Teachers of Mathematics. *Changing School Mathematics: A Responsive Process*. Reston, Va.: The Council, 1981.

_____. *Handbook for Conducting Equity Activities in Mathematics Education*. Reston, Va.: The Council, 1984.

_____. *An In-Service Handbook for Mathematics Education*. Reston, Va.: The Council, 1977.

National Staff Development Council. *Conducting Effective Staff Development Programs*. Oxford, Ohio: 1985.

New Jersey State Department of Education. *Improving Mathematical Problem- solving Skills in the Middle Grades: A Staff Training Manual*. Trenton, N.J.: The Department, 1985.

Showers, Beverly. "Teachers Coaching Teachers." *Educational Leadership*, April 1985.

Sparks, Georgea Mohlman. "Synthesis of Research on Staff Development for Effective Teaching." *Educational Leadership*, November 1983.

Wood, Fred H., et al. "Practitioners and Professor Ogra on Effective Staff Development Practices." *Educational Leadership*, October 1982.